FRUITS YOU LOVE TO EAT

PEACHES

AMY CULLIFORD

A Crabtree Roots Book

Crabtree Publishing
crabtreebooks.com

School-to-Home Support for Caregivers and Teachers

This book helps children grow by letting them practice reading. Here are a few guiding questions to help the reader with building his or her comprehension skills. Possible answers appear here in red.

Before Reading:

- What do I think this book is about?
    - *I think this book is about how good peaches taste.*
    - *I think this book will tell me how peaches are grown.*
- What do I want to learn about this topic?
    - *I want to learn why peaches are fuzzy on the outside.*
    - *I want to learn why peaches have a pit in the middle.*

During Reading:

- I wonder why...
    - *I wonder why peaches are juicy.*
    - *I wonder why peaches are only found in stores at certain times of the year.*
- What have I learned so far?
    - *I have learned that peaches are fruits.*
    - *I have learned that peaches grow on trees.*

After Reading:

- What details did I learn about this topic?
    - *I have learned that peach trees have flowers that turn into peaches.*
    - *I have learned that peaches can be white, yellow, red, or orange.*
- Read the book again and look for the vocabulary words.
    - *I see the word **pits** on page 10 and the word **fuzz** on page 12. The other vocabulary words are found on page 14.*

Peaches are **fruits.**

All peaches grow on peach **trees**.

The trees have **flowers** on them that turn into peaches.

Peaches can be white, yellow, red, or orange.

All peaches have big brown **pits**.

Peaches also have peach **fuzz**.

Word List

Sight Words

all	brown	on	the
also	can	or	them
are	grow	orange	turn
be	have	red	white
big	into	that	yellow

Words to Know

flowers

fruits

fuzz

peaches

pits

trees

38 Words

Peaches are **fruits**.

All peaches grow on peach **trees**.

The trees have **flowers** on them that turn into peaches.

Peaches can be white, yellow, red, or orange.

All peaches have big brown **pits**.

Peaches also have peach **fuzz**.

FRUITS YOU LOVE TO EAT
PEACHES

Written by: Amy Culliford
Designed by: Rhea Wallace
Series Development: James Earley
Proofreader: Melissa Boyce
Educational Consultant: Marie Lemke M.Ed.

Photographs:
Shutterstock: Natalay Studio: cover; Tum UR: p. 1; Maria Dry Shout: p. 3; Nyura: p. 5; Kostyantyn Sulima: p. 7; MK Studio: p. 9; fotocat5: p. 11; Lord Benedict: p. 13

Crabtree Publishing

crabtreebooks.com 800-387-7650

In Canada: We acknowledge the financial support of the Government of Canada through the Canada Book Fund for our publishing activities.

Printed in Canada/122023/20231201

Published in Canada
Crabtree Publishing
616 Welland Ave.
St. Catharines, Ontario
L2M 5V6

Published in the United States
Crabtree Publishing
347 Fifth Ave
Suite 1402-145
New York, NY 10016

Library and Archives Canada Cataloguing in Publication
Available at Library and Archives Canada

Library of Congress Cataloging-in-Publication Data
Available at the Library of Congress

Hardcover: 978-1-0398-0977-2
Paperback: 978-1-0398-1030-3
Ebook (pdf): 978-1-0398-1136-2
Epub: 978-1-0398-1083-9